Nuclear Engineering Division

Held at the 2023 AIChE Annual Meeting

Orlando, Florida, USA
5-10 November 2023

ISBN: 978-1-7138-9289-2

Printed from e-media with permission by:

Curran Associates, Inc.
57 Morehouse Lane
Red Hook, NY 12571

Some format issues inherent in the e-media version may also appear in this print version.

Printed with permission by Curran Associates, Inc. (2024)

For permission requests, please contact AIChE
at the address below.

AIChE
120 Wall Street, FL 23
New York, NY 10005-4020

Phone: (800) 242-4363
Fax: (203) 775-5177

www.aiche.org

Additional copies of this publication are available from:

Curran Associates, Inc.
57 Morehouse Lane
Red Hook, NY 12571 USA
Phone: 845-758-0400
Fax: 845-758-2633
Email: curran@proceedings.com
Web: www.proceedings.com

TABLE OF CONTENTS

ENVIRONMENTAL ADVANCE IN NUCLEAR AND HAZARDOUS WASTE PROCESSING AND DISPOSAL

THEORY, MODELING, AND SIMULATION OF NUCLEAR CHEMICAL PROCESSES

THEORY, MODELING, AND SIMULATION OF MATERIALS IN FUSION PLASMA

162a Reducing Mercury Emissions and Increasing Mercury Recovery from Processing High Level Radioactive Waste in the Defense Waste Processing Facility at the Savannah River Site

Dan Lambert, Anthony Howe, Sean Noble

The Defense Waste Processing Facility (DWPF) at the Savannah River Site (SRS) near Aiken, South Carolina, processes and immobilizes the legacy waste from the production of nuclear weapons materials. Although great progress has been made in processing and vitrifying high level waste (HLW) at SRS, approximately 100 million gallons of radioactive waste remains to be processed.

In this paper, we discuss a flowsheet alternative that is being considered for use in processing high level radioactive waste to increase condenser efficiency, which could potentially decrease mercury emissions from the DWPF stack. The flowsheet change aims to further decrease the air purge during chemical processing. The lower purge is possible due to the elimination of hydrogen, ammonia and three flammable decomposition gases as a result of replacement of the antifoam and the use of glycolic acid instead of formic acid.

Testing demonstrated that the lower air purge is feasible. However, the lower purge also decreased detection limits for offgas species and allowed the identification of carbon monoxide (CO) for the first time. CO is produced concurrently with carbon dioxide (CO_2). In addition, one of the tests used nitrogen as a purge gas in place of air and this led to almost complete recover of mercury during the DWPF simulation. This new discovery offers a potential improvement to recovering mercury during processing. At present, no mercury is recovered in DWPF as planned. The potential for this lower air purge both increasing the condenser efficiency, decreasing mercury release through the stack and increasing mercury recovery will be discussed.

162b Hanford Sludge Washing for Aluminum and Phosphate Dissolution

Amy Westesen, Abigail Robb, Nicholas Cappella, Christian Alvarez, Reid Peterson, Nicholas Machara

The nuclear waste at the Department of Energy's (DOE) Hanford site represents one of the single largest environmental cleanup efforts currently underway. Large volumes of highly radioactive waste, generated during past plutonium production operations, are stored in underground tanks on the Hanford site. Effective disposition of this waste will require processing the sludge to immobilize the waste via vitrification into a stable glass form for long-term disposal. Conceptual flowsheets have proposed settling of the solids in process vessels as a primary step in providing sludge to the vitrification facility. Wastes with high concentrations of aluminum and phosphate have been identified as particularly challenging to process and simultaneously make up the majority of insoluble components in the sludge. This presentation details low temperature dissolution studies of aluminum and phosphate sludge simulants in a variety of wash matrices mimicking potential retrieval operations. Slurry compositions were prepared to match difficult layers in sludge tank AN-106 in order to bound high phosphate/high aluminum operating conditions experienced during processing this tank. Empirical testing data were compared to modeled solubility limits to determine dissolution rates under various wash conditions as well as demonstrate sludge volume reduction capabilities. Removal of these components can significantly reduce (2x) the volume of high-level waste glass produced, decrease operating costs, and improve processing performance.

162c Rheology of Concentrated Nuclear Waste Suspensions: Reduced Yield Stress By Additive Blending

Olivia Pickup, Alexander Lockwood, Martyn Barnes, Timothy N. Hunter, David Harbottle

A significant activity of the UK nuclear sector is decommissioning, which involves removing and packaging legacy wastes for their long-term safe storage. Intermediate level waste materials include magnesium alloy fuel cladding from Magnox reactors that has corroded over time in underwater storage to form a highly complex suspension. The aim is to encapsulate the suspension waste in cement, but to do this there is a compromise between the suspension rheology and total waste volume, i.e. to reduce total waste volume the solids content should be high. However, higher solids content produces undesirable rheological properties such as high yield stress and shear thickening. To mitigate these issues, the current research has explored the use of nano-silica (SiO_2), an inorganic additive, to lower the yield stress of concentrated magnesium hydroxide ($Mg(OH)_2$) suspensions.

The yield stress of 27 vol% $Mg(OH)_2$ suspension was 75 Pa, but when blended with 3 vol% SiO_2, so that the total solids content was 30 vol%, the yield stress decreased by 40% (Fig. 1a). Samples from the blended suspensions were imaged by scanning electron microscopy (SEM) and x-ray computed tomography (X-ray CT) (Figs. 1b and c) to determine the distribution and dispersion of the nano-silica throughout the $Mg(OH)_2$ network. Initially, good distribution but poor dispersion of SiO_2 was seen with large amounts of particle clustering. The drop in yield stress is attributed to the distribution of SiO_2 which modifies the contact mechanics between $Mg(OH)_2$ aggregates and promotes a lubrication effect. With aging, the yield stress increased to >200 Pa after 7 days (Fig. 1a) at ambient conditions, with the onset for yield stress growth occurring after 1 day. This gradual stiffening of the particle network was attributed to the slow dissolution of SiO_2 and its reaction with soluble Mg^{2+} ions forming a magnesium silicate hydrate precipitate between particles, as seen in high-resolution SEM images.

The study demonstrates the use of nano-additives to lower the yield stress of concentrated suspensions and explores the timescales of physical lubrication in competition with the onset of chemical gelation. Practical application of the method can lead to significant cost savings for the UK nuclear decommissioning program by reducing the total number of waste packages needed for long-term storage.

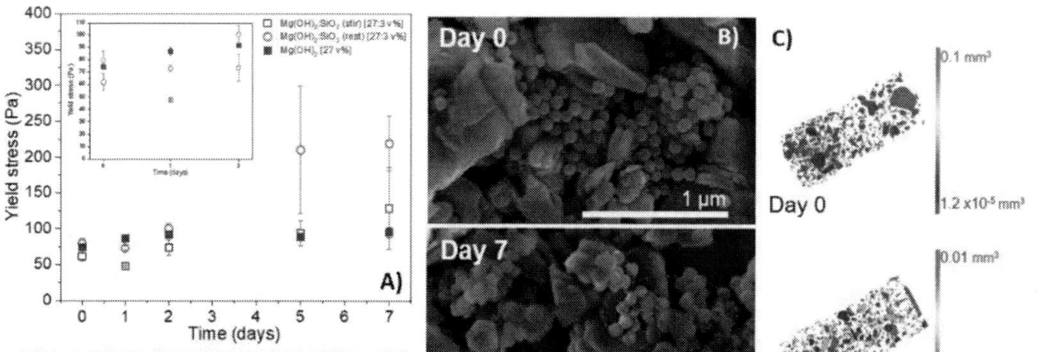

Figure 1: A) Yield stress reduction and growth observed in $Mg(OH)_2 + SiO_2$ [27:3 v%] over 7 days, B) scanning electron micrographs indicating dissolution of SiO_2 and C) the comparison of SiO_2 cluster size over time with X-ray CT.

363d Fractional Neutron Point Kinetic Modeling of the UMass Lowell Research Reactor

Kashief Bogannam, Sukesh K. Aghara, Valmor F. de Almeida

Fractional derivative order modeling has been the subject of much study for many physical processes in recent years. One such process is the diffusion of neutrons in a nuclear reactor, which can be described as a sub-diffusive process when the underlying qualifiers of the diffusion assumption are challenged. This has led to the development of a fractional neutron point kinetics (FNPK) model for a nuclear reactor. This work provides an analysis of the neutron density response of the isothermal FNPK model, as well as the FNPK for the UMass Lowell research reactor (UMLRR). Several scenarios of the neutron source term as well as varying reactivity insertion are tested on the isothermal model, and analysis on the effects of the fractional parameters κ and τ^{κ} on predicted neutron density is presented. A detrended fluctuation analysis is performed on UMLRR experiment data, producing the applicable κ values for modeling of the reactor. Finally, a designated UMLRR experiment is simulated using the FNPK coupled with two-temperature, two-poison feedback for several τ^{κ} values to determine the applicability of the FNPK model.

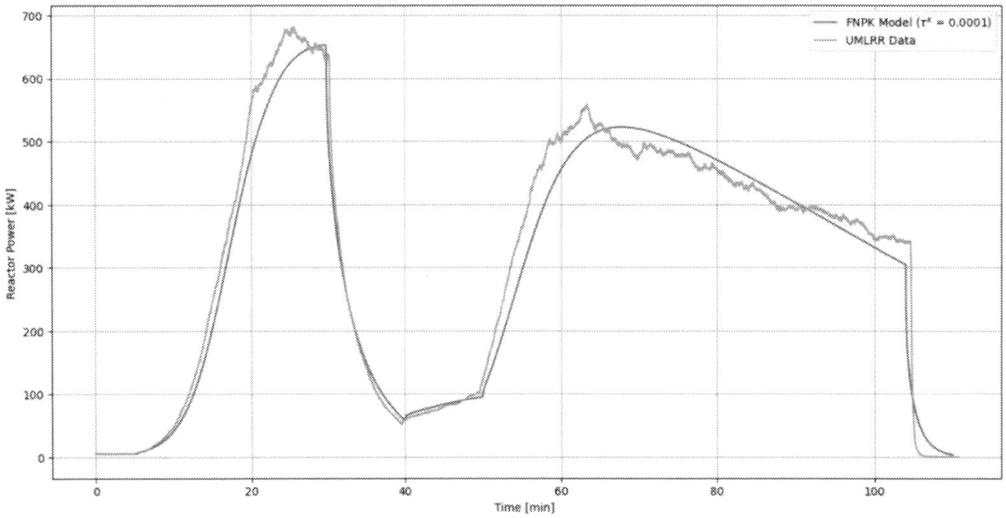

363e Solvent-Controlled Ligand Conformational Energetics in Liquid-Liquid Extraction: A Combined Simulation and Experimental Study on Molecular Solvents and Ionic Liquids

Xiaoyu Wang, Srikanth Nayak, Richard Wilson, Lynda Soderholm, Michael Servis

Solvent extraction is the most widely used technique in recovering radioactive metals and reprocessing aqueous phase of nuclear fuels. Designing better extractants in solvent extractions is a research frontier in metal ion separations that typically focuses on the direct extractant-metal interactions. However, a more detailed understanding of energetic drivers of separations beyond primary metal coordination is often lacking, especially including the role of solvent in the extractant phase. In this work, we propose a new mechanism for enhancing metal extraction energetics with nanostructured solvents. Using molecular dynamics simulations to estimate free energies, we find that the organic solvent can reshape the energetics of the extractant's intramolecular conformational landscape. Our investigations have covered some multidentate extractants, including malonamides and CMPO, in different types of solvents, from traditional molecular solvents such as dodecane and TBP, to state-of-the-art ionic liquids (IL). By promoting reorganization of the extractant molecule into its binding conformation, our findings reveal how particular solvents can ameliorate this unfavorable step of the metal extraction process. In particular, the charge alternating nanodomains formed in ILs substantially reduce the free energy penalty associated with extractant reorganization. These findings provide insights into the potential energy drivers underpinning the notable separation improvements reported with ILs and suggest a new approach to designing effective separations using a molecular-level understanding of the solvent effects.

363f Direct Parameter Optimization for Chemical Reaction Kinetics

Valmor de Almeida, Jack J. F. Kennedy

A traditional method for calibration of parameters of chemical reaction rate laws from experimental concentration time-dependent data involves an ODE-solver in the loop. This is a laborious algorithm that relies on the solution of the ODE system at each optimization iteration step. Since ODE equations can be very sensitive to parameters and a solution can become unstable, this method is not without challenges. Alternatively, this work experiments with a direct method of parameter optimization using the time-derivative of the experimental data directly. This avoids the use of an ODE solver in the parameter optimization loop but requires differentiation of the experimental data. We use a smooth approximation of the concentration data that allows for differentiation, and solve the non-linear algebraic parameter optimization problem via a least-squares residual method. We apply this method to contrived reaction mechanisms analyzing its performance as compared to golden values. We investigate the role of rank-deficiency of the normal equations of the least-squares Gauss-Newton's method, and demonstrate the usage of the Hessian matrix for the general reaction rate power-law form. The eigenvalues of the Hessian matrix provide useful information about the stationary point of the objective function converged to. Last, computational experiment are made for the radiolysis of water reaction mechanism with and without borated water. In general, we find parameter optimization of reaction mechanisms for water radiolysis of interest to nuclear reactors and chemical reprocessing to be difficult to calibrate accurately. Also, the literature seems lacking on how reaction mechanism have been calibrated against experimental data which puts in question the applicability/trasferability of the reaction rate formulae.

Analysis of Scoop System in a Mechanically Driven Gas Centrifuge

Sahadev Pradhan

Chemical Technology Division, Bhabha Atomic Research Centre, Mumbai-400 085, India.

ABSTRACT:

In this study we investigate the performance of a stationary scoop system (scoop tip and its arm) in a mechanically driven gas centrifuge at wall Mach number Ma_{wall} in the range 4 to 8, wall pressure P_{wall} in the range 20 to 100 m-bar, scoop arm radius of curvature (extremity radius) R_{roc} in the range 5 to 30 mm, scoop wall gap d_{wall} in the range 5 to 20 mm, and slenderness ratio (major to minor axis) of the elliptic scoop in the range 1.2 to 6 [1, 2, 3, 4, 5, 6, 7, 8, 9, 10]. The analytical model is formulated based on the steady state assumption where the accelerating moment exerted by the rotating wall and bottom end cap on the rotating gas contained in it, is balanced by the decelerating moment due to the aerodynamic resistance of the scoop tip and its arm, which allow us to evaluate the scoop plane Mach number as a function of wall Mach number with the parameter $A\ Kn_{axis}$. Here, Kn_{axis} is the Knudsen number at the axis of the rotating cylinder, and the factor "A", characterized a particular scoop system (shape and size of the scoop tip and its arm). In a high speed rotating field we examine the stagnation to axis pressure ratio as a function of wall Mach number for different scoop systems and the result indicates that scoop system having smaller dimension exhibit higher pressure recovery with modest slow-down of the circumferential velocity of the rotating gas. An important finding is that the scoop arm with higher radius of curvature (tending towards more circular shape) exhibit less decelerating moment and generates a reduced amount of secondary radial flow. Therefore, in order to enhance the secondary radial flow towards the axis, which further excites the secondary axial flow, which could be very important for the centrifugal gas separation processes, scoop arm with lower radius of curvature (extremity radius) is preferred ((Pradhan & Kumaran, *J. Fluid Mech.*, vol. 686, 2011, pp. 109-159); (Kumaran & Pradhan, *J. Fluid Mech.*, vol. 753, 2014, pp. 307-359)). The analysis shows that with the increase of wall pressure from 20 to 100 m-bar the decelerating moment of the scoop-tip and its arm increases, which results more slow-down of the circumferential velocity of the rotating gas at the scoop plane. The analysis also indicates that at a given wall Mach number, scoop tip exhibit more decelerating moment compared to the scoop-arm having same dimensions, and with the increase of wall pressure the difference between them reduces. Therefore, the required magnitude of deceleration can be achieved at a given wall pressure through proper combination of the scoop tip and its arm dimension. The effective scoop arm length that provides most of the scoop arm deceleration, is estimated with respect to the total arm length, and the result shows that arm-profile of the effective length portion with various winged shape is an important design aspect while operating at high wall pressure.

Keywords: High-speed rotating flow, Scoop system, Analytical model, Accelerating and decelerating moments, Scoop plane Mach number.

References:

[1] PRADHAN, S. & KUMARAN, V. 2011 The generalized Onsager model for the secondary flow in a high-speed rotating cylinder. *J. Fluid Mech.* 686, 109.

[2] KUMARAN, V & PRADHAN, S. 2014 The generalized Onsager model for a binary gas mixture. *J. Fluid Mech.* 753, 307.

[3] WOOD, H.G. & MORTON, J.B. 1980 Onsager's pancake approximation for the fluid dynamics of a gas centrifuge. *J. Fluid Mech.* 101, 1.

[4] WOOD, H. G. & SANDERS, G. 1983 Rotating compressible flows with internal sources and sinks. *J. Fluid Mech.* 127, 299.

[5] WOOD, H. G. & BABARSKY, R. J. 1992 Analysis of a rapidly rotating gas in a pie-shaped cylinder. J. Fluid Mech. 239, 249.

[6] OLANDER, D. R. 1981 The theory of uranium enrichment by the gas centrifuge. *Prog. Nucl. Energy* 8, 1.

[7] ZIPPE G. 1960 The development of short bowl ultracentrifuges, University of Virginia, Report no: EP-4420-101-600.

[8] Napolitano L. G., Monti R., Losito U.,*Studio del sistema di estrazione del gas di un'ultracentrifuga*, L'Aerotecnica Missili e Spazio n. 3, 1973.

[9] VINCENTI. 1967 Introduction to Physical Gas Dynamics. Wiley, New York.

[10] CHAPMAN, S. & COWLING, T. G. 1970 The Mathematical Theory of Non-Uniform Gases, 2nd edition. Cambridge, Cambridge University Press.

323a On the Onset of 'Fuzz' Formation in Plasma-Facing Materials: A Hierarchical Multiscale Modeling Approach

Dwaipayan Dasgupta, Asanka Weerasinghe, Dimitrios Maroudas, Brian D. Wirth

Tungsten (W) and tungsten alloys are widely viewed as the most promising plasma-facing material (PFM) candidates for divertor and first-wall systems in nuclear fusion reactors. However, experiments aiming to investigate PFM performance under continuous operation and high first-wall temperature have shown that a nanostructure with a fuzz-like morphology develops on the plasma-exposed W surface under the operating conditions of temperature, helium (He) impact energy, and He flux expected for ITER's divertor. Formation of such fuzz surface nanostructure in tungsten may adversely influence the reactor performance and operation.

Fuzz formation in plasma-facing component (PFC) tungsten is a complex multi-physics phenomenon resulting from the combined effects of processes that include driven surface diffusion, subsurface He bubble dynamics, He bubble bursting, and dislocation loop punching, as well as material property anisotropies and changes in material thermophysical properties in the damaged tungsten. The development of a predictive model capable of capturing the formation and evolution of the fuzz-like complex surface morphology mediated by the various underlying dynamical processes that are characterized by disparate spatiotemporal scales is particularly challenging. While large-scale molecular-dynamics (MD) simulations have been used to successfully capture and explore many of the aforementioned phenomena [1], simulating the experimentally relevant spatiotemporal scales (mm-hr) for surface nanostructure formation in PFC tungsten is well beyond the scope of such atomistic simulations. Therefore, we followed a hierarchical multiscale modeling paradigm to develop a continuum-scale model for simulating the onset of fuzz formation in He plasma-irradiated tungsten [2].

In our modeling framework, large-scale MD simulation results are used to parameterize the constitutive equations required for the closure of the continuum-scale model for the surface morphological response of the plasma-facing material and targeted MD simulations are used to determine the thermophysical properties [3, 4] of He-implanted tungsten. Based on this atomistically-informed continuum-scale model, we have conducted self-consistent numerical simulations of the He-implanted tungsten surface morphological evolution and validated the model by comparing the simulation predictions with measurements from carefully designed experiments [2]. Furthermore, we have explored the effects on the PFC surface morphology and growth kinetics of the surface temperature [5], the elastic softening of the near-surface PFC region [6], the formation of nanometer-scale holes on the PFC surface due to He bubble bursting [7], and the formation of surface adatom-vacancy pairs under low-energy He implantation [8]. The simulation predictions are compared with experimental data and provide a fundamental interpretation to experimental observations.

References: [1] K. D. Hammond, S. Blondel, L. Hu, D. Maroudas, and B. D. Wirth, Acta Mater. **144**, 561-578 (2018); [2] D. Dasgupta, R. D. Kolasinski, R. W. Friddle, L. Du, D. Maroudas, and B. D. Wirth, Nucl. Fusion **59**, 086057 (2019); [3] A. Weerasinghe, B. D. Wirth, and D. Maroudas, ACS Appl. Mater. Interfaces **12**, 22287 (2020); [4] A. Weerasinghe, B. D. Wirth, and D. Maroudas, J. Appl. Phys. **132**, 185102 (2022); [5] D. Dasgupta, D. Maroudas, and B. D. Wirth, Surf. Sci. **698**, 121614 (2020); [6] C.-S. Chen, D. Dasgupta, A. Weerasinghe, B. D. Wirth, and D. Maroudas, Nucl. Fusion **61**, 016016 (2021); [7] C.-S. Chen, D. Dasgupta, B. D. Wirth, and D. Maroudas, J. Appl. Phys. **129**, 193302 (2021); [8] C.-S. Chen, D. Dasgupta, A. Weerasinghe, K. D. Hammond, B. D. Wirth, and D. Maroudas, Nucl. Fusion **63**, 026033 (2023).

323b Thermomechanical Properties of Plasma-Facing Tungsten

Asanka Weerasinghe, Brian D. Wirth, Dimitrios Maroudas

Investigating the impact of helium (He) ion implantation on the thermomechanical properties of tungsten is of utmost importance for evaluating tungsten as a plasma-facing component (PFC) in nuclear fusion devices. To this end, here, we report results on the thermomechanical behavior of single-crystalline tungsten containing structural defects, voids and over-pressurized He nanobubbles, related to the conditions of plasma exposure in nuclear fusion devices. We develop PFC models as regular distributions of spherical nanoscale-size He bubbles in the tungsten matrix to investigate and predict the thermomechanical response under realistic plasma exposure conditions. Based on molecular-dynamics (MD) simulations using a properly parameterized machine learning Spectral Neighbor Analysis Potential (SNAP), we characterize the response of tungsten to plasma exposure and determine the elastic properties, tensile yield strength, and thermal expansion coefficient of PFC tungsten as a function of its porosity and He content.

We find that the Young modulus of PFC tungsten softens substantially as a function of porosity, following an exponential scaling relation, in addition to the softening caused in the tungsten matrix with increasing temperature [1]. The presence of these nanoscale spherical defects reduces the yield strength of tungsten in a monotonically decreasing fashion, obeying an exponential scaling relation as a function of tungsten matrix porosity and He concentration [1]. Detailed analysis of the structural response of PFC tungsten near the yield point reveals that yielding is initiated by emission of dislocation loops from bubble/matrix interfaces followed by gliding and growth of these loops and dislocation reactions. Moreover, dislocation gliding on twin systems nucleates twin regions in the tungsten matrix. These dynamical processes reduce the stress in the matrix substantially [1].

We also distinguish between two approaches of filling the bubbles with He, where the amount of He in the bubble can or cannot vary with temperature. In the former case, the thermal expansion coefficient decreases monotonically with increasing the porosity and He content of the tungsten matrix, while in the latter case, the thermal expansivity increases monotonically with increasing porosity and He content [2]. The latter condition, where the He content in the bubble is determined at the implantation temperature and remains constant with varying temperature in the tungsten matrix, is consistent with He species transport in PFC tungsten and implies the development of biaxial compressive thermal strains in the PFC material that contribute to accelerating the growth of a nanostructure on PFC tungsten surfaces.

Our analyses advance the fundamental understanding of thermomechanical properties in PFC tungsten and contributes to the development of a thermophysical property database for properly incorporating effects of realistic particle and heat loads into modeling the dynamical response of PFC tungsten under fusion reactor operating conditions.

[1] A. Weerasinghe, E. Martinez, B. D. Wirth, and D. Maroudas, *ACS Appl. Mater. Interfaces* **15**, 8709–8722 (2023).

[2] A. Weerasinghe, B. D. Wirth, and D. Maroudas, *J. Appl. Phys.* **132**, 185102 (2022).

323c Helium Bubble Size Effects on the Surface Morphological Response of Plasma-Facing Tungsten

Omeet Patel, Dwaipayan Dasgupta, Brian D. Wirth, Dimitrios Maroudas

Tungsten (W) is the chosen plasma-facing component (PFC) material for the divertor in the International Thermonuclear Experimental Reactor (ITER) due to its superior properties, including exceptional thermomechanical properties and low sputtering yield. Numerous experimental studies have established that PFC tungsten suffers severe surface degradation as a result of exposure to high fluxes of helium (He) and extremely high heat loads. Specifically, high fluxes of low-energy helium ions implanted in tungsten within the temperature range from 900 K to 2000 K are responsible for the formation of a "fuzz"-like surface nanostructure, which consists of fragile nanoscale-sized crystalline tendrils. This nanotendril formation is driven by stress-induced surface atomic diffusion, with stress originating from the over-pressurized He bubbles formed due to He implantation in the near-surface region of PFC tungsten, and has adverse effects on the mechanical behavior and structural response of PFC tungsten as well as on the reactor performance. Based on several experimental observations [1, 2], in our previous studies on surface morphological evolution of PFC tungsten, we considered the average He bubble radius to be in the range of 0.75 nm to 1.25 nm. However, recent experimental findings [3] have revealed a broader size range of He bubbles formed in the near-surface region of PFC tungsten as a result of He implantation with average bubble radii extending up to 3 nm.

Motivated by these recent experimental results, we report here a simulation study on the effects of He bubble size on the surface morphological evolution and pattern formation of PFC tungsten. Our analysis focuses on the W(110) surface orientation and is based on self-consistent dynamical simulations according to an atomistically-informed, continuum-scale surface evolution model that has been developed following a hierarchical multiscale modeling strategy and can access the spatiotemporal scales of relevance to fuzz formation. The model accounts for PFC surface diffusion driven by the biaxial compressive stress originating from the over-pressurized helium bubbles in a thin nanobubble layer, which forms in the near-surface region of PFC tungsten as a result of He implantation, in conjunction with formation of self-interstitial atoms in tungsten that diffuse toward the surface. The model also accounts for the flux of surface adatoms generated as a result of surface vacancy-adatom pair formation upon He implantation, which contributes to the anisotropic growth of surface nanostructural features due to the different rates of adatom diffusion along and across step edges of islands on the tungsten surface [4].

Using the model described above, we find that varying the average He bubble radius has a direct impact on the kinetics of the surface morphological response of PFC tungsten and the resulting surface topography. Specifically, increasing the He bubble size leads to a decrease in the growth rate of the nanotendrils emanating from the PFC surface, which is interpreted based on the corresponding decrease in the stress level in the tungsten nanobubble region. Moreover, the separation distance between the resulting surface features increases with an increase in the average He bubble size. We analyze this coarsening effect systematically and find that it is a thermally activated process with the average surface feature separation depending on temperature according to an Arrhenius relation. The results of our computational analysis are in good agreement with theoretical predictions based on linear stability analysis of the PFC surface morphological response.

[1] K. Wang, R. P. Doerner, M. J. Baldwin, F. W. Meyer, M. E. Bannister, A. Darbal, R. Stroud, and C. M. Parish, Sci. Rep. **7**, 42315 (2017).

[2] M. Thompson, P. Kluth, R. P. Doerner, N. Kirby, and C. Corr, Nucl. Fusion **55**, 042001 (2015).

[3] M. Ialovega, E. Bernard, M. Barthe, R. Bisson, A. Campos, M. Cabie, T. Neisus, R. Sakamoto, A. Kreter, C. Grisolia, T. Angot, and C. Martin, Nucl. Fusion **62**, 126022 (2022).

[4] C.-S. Chen, D. Dasgupta, A. Weerasinghe, K. D. Hammond, B. D. Wirth, and D. Maroudas, Nucl. Fusion **63**, 026033 (2023).

323d Impact of Soret Effect on Hydrogen and Helium Retention in Plasma-Facing Tungsten Under ELM-like Conditions

Dimitrios Maroudas, Dwaipayan Dasgupta, Sophie Blondel, Brian D. Wirth, Enrique Martinez

Plasma-facing components (PFCs) in nuclear fusion reactors are expected to be exposed to extremely high particle fluxes and heat loads that modify the PFC materials microstructure. These fluxes will create strong gradients of temperature and concentration of diverse species, including He and H atoms and small mobile helium clusters. In addition to these impurity species, intrinsic defects, such as self-interstitial atoms (SIAs), also are generated in the PFC near-surface region. The aforementioned thermal and concentration gradients will strongly affect the migration of these defects and impurity species in the PFC material.

To fundamentally understand such effects, we have used nonequilibrium molecular-dynamics (NEMD) simulations to analyze the transport of He, mobile helium clusters, H, and SIAs in the presence of a thermal gradient in tungsten, the chosen PFC material for the divertor of ITER (International Thermonuclear Experimental Reactor). We have found that all the species examined tend to migrate toward the hot regions of the tungsten sample [1]. The resulting species concentration profiles are exponential distributions in the direction of the imposed temperature gradient, rising toward the hot regions of the sample, in agreement with irreversible thermodynamics analysis [1]. For all the species examined, intrinsic point defects and impurities, we have found that the resulting species flux is directed opposite to the heat flux, indicating that species transport is governed by a Soret effect, namely, thermal-gradient-driven diffusion, characterized by a negative heat of transport that drives species transport uphill, *i.e.*, from the cooler to the hot regions of the tungsten sample [1,2].

The findings of our thermal and species transport analysis have been implemented in our cluster-dynamics code, Xolotl, which has been used to compute temperature and species profiles over spatiotemporal scales representative of PFC tungsten under typical reactor operating conditions, including extreme heat loads at the plasma-facing surface characteristic of plasma instabilities that induce edge localized modes (ELMs) [2]. We demonstrate that the steady-state species profiles, when properly accounting for the Soret effect, vary significantly from those where temperature-gradient-driven transport is not accounted for and discuss the implications of such a Soret effect on the response to plasma exposure of plasma-facing tungsten. Although our cluster-dynamics simulations do not yet include self-clustering of helium or hydrogen blister formation, our simulation results show that the Soret effect substantially reduces helium and hydrogenic species retention inside PFC tungsten [2].

[1] E. Martinez, N. Mathew, D. Perez, S. Blondel, D. Dasgupta, B. D. Wirth, and D. Maroudas, *J. Appl. Phys.* **130**, 215904 (2021).

[2] D. Dasgupta, S. Blondel, E. Martinez, D. Maroudas, and B. D. Wirth, *Nucl. Fusion*, under review (2023).

323e Empirical Model of Helium Bubble Growth and Bursting Near W{100} Surfaces and Equation of State for Helium Bubbles in Tungsten

Wahyu Setyawan, Dwaipayan Dasgupta, Sophie Blondel, Giridhar Nandipati, Karl Hammond, Dimitrios Maroudas, Brian D. Wirth

Fuzz formation in helium-implanted tungsten (W) is a complex, multi-physics phenomenon resulting from the combined effects of processes that include driven surface diffusion, subsurface helium (He) bubble dynamics, He bubble bursting, and loop punching, as well as anisotropies of material properties and substantial changes in material thermophysical properties in the damaged tungsten as a function of He content. Here, we study the subsurface bubble dynamics in order to develop predictive capabilities for He content by focusing on three important physical phenomena and physical dependences, namely, subsurface bubble growth, bubble bursting, and the effect of temperature on subsurface bubble dynamics.

We have performed molecular-dynamics (MD) simulations to study the growth of He bubbles under a W(100) surface until the bubbles burst. We study bubble growth as a function of the initial nucleation depth of the bubbles. During growth, successive loop-punching events are observed, accompanied by shifts in the depth of the bubble toward the surface. Subsequently, the MD data are used to derive empirical models that describe the conditions that cause the loop punching and bursting events. Simulations have been performed at 500, 933, 1500, 2000, and 2500 K to fit the respective parameters in the models. In the loop punching model, for bubbles consisting of N_V vacancies and N_{He} helium atoms, the N_{He}/N_V ratio that causes the event, the resulting increase in N_V, and the associated shift of the bubble depth are formulated as a function of N_V and temperature, T. In the bursting model, a bubble must reach a certain depth and N_{He}/N_V ratio, simultaneously, in order to burst. The burst depth and N_{He}/N_V ratio are also modeled as a function of N_V and T. To compute the pressure in the bubble at the loop punching and bursting events from the models, we derive an equation of state (EOS) for He bubbles in tungsten. The EOS is fitted to data from MD simulations of He bubbles in bulk W that span a wide range of gas density and bubble sizes up to about 3 nm in diameter. For completeness, a model is also derived to compute the bubble volume for a given N_V, N_{He}/N_V ratio, and T. The pressure of subsurface bubbles at the loop punching events as calculated using the bubble-EOS and the volume model agrees well with the pressure obtained directly from the MD simulations. The majority of the loop punching events occur at bubble pressures between 20 and 60 GPa, depending on the bubble size and temperature. The larger the bubble and the higher the temperature, the lower the bubble pressure. Furthermore, we find that at a higher temperature, a bubble can burst from a deeper region.

In addition, a new EOS for a free helium gas is derived to improve an existing free-gas EOS that was fitted to experimental data over the range of 0.2–2 GPa and 75–300 K. The existing EOS agrees well with our MD data up to 500 K. However, it overpredicts the bubble pressure at temperatures higher than 933 K, and becomes increasingly inaccurate at higher temperatures and pressures higher than 2 GPa. The new free-gas EOS can accurately predict all the MD data included in the analysis (which span up to 54 GPa at 2500 K).

323f Helium Bubble Coalescence in Plasma-Facing Tungsten

Asanka Weerasinghe, Brian D. Wirth, Dimitrios Maroudas

Predicting surface nanostructure formation on tungsten (W) under plasma exposure conditions is of utmost importance when tungsten is used as a plasma-facing component (PFC) in nuclear fusion devices. Subsurface helium (He) transport and He bubble dynamical processes, such as bubble coalescence and bubble bursting, play crucial roles in the near-surface structural evolution and surface morphological response of the PFC material. However, in spite of its importance to the dynamical response of the PFC material, a fundamental understanding of the He bubble coalescence mechanism in PFC tungsten remains incomplete. To this end, here, we report results of a systematic study of He bubble coalescence in PFC tungsten based on atomistic modeling and simulations. Our PFC tungsten models consist of He nanobubbles embedded in the tungsten matrix with bubble sizes and He density relevant to the conditions of plasma exposure in nuclear fusion devices. Based on molecular-statics (MS) and molecular-dynamics (MD) simulations, we explore the governing thermodynamics of bubble coalescence and its associated parameter space, with parameters that include bubble size, bubble separation distance, bubble pressure, bubble growth rate, and bubble location (near the PFC surface or in the bulk material), and design targeted dynamical simulations to study the underlying kinetics and characterize the governing coalescence mechanisms.

We find that the interaction energetics between two He bubbles in W can be described as an elastic interaction perturbation to a finite-width square-well potential. The width of the square-well potential has a direct correlation to the bubble pressure and corresponds to the capture radius of the two-bubble system, beyond which the bubbles interact strongly to facilitate their coalescence. In general, smaller bubbles tend to be captured by the larger bubbles due to the higher bubble pressure (He/Vacancy ratio) in the smaller bubbles. When the two bubbles are sufficiently close to each other, the defective W regions around the spherical bubble surfaces merge at the narrow W gap that separates the two He bubbles forming a defective region with a characteristic "peanut" or dumbbell shape. Furthermore, we find that when the W gap between the two He bubbles narrows down to about one W layer, the W atoms in that layer are pushed away from each other to create a channel between the two bubbles, triggering the coalescence mechanism that allows He atoms to migrate between the two bubbles through a stress-driven interdiffusion process. This mechanism is aided by the formation of Frenkel pairs; the channel opening consists of W vacancies with the corresponding self-interstitial atoms occupying sites near the channel. Moreover, continuing He implantation into the bubbles increases the pressure of the post-coalescence bubble configuration, which causes emissions of ½<111> and <100> dislocation loops/segments from the bubble surface. We also find that when the two He bubbles are sufficiently large and far apart from each other (at a distance larger than the capture radius), dislocation emissions between the two bubbles cause a narrowing of the gap between them, thus accelerating the bubble coalescence mechanism.

Our analyses advance our fundamental understanding of helium bubble coalescence in PFC tungsten and will contribute to further improve the predictive capabilities of coarse-grained models of near-surface structural evolution and surface morphological evolution of PFC tungsten under fusion reactor operating conditions.

289b Transient Behavior of Dead-End Filtration at Low Solids Concentration

Philip Schonewill, Jarrod Allred, John Geeting

The Hanford Site's Direct Feed Low-Activity Waste process performs a pretreatment step prior to planned vitrification in the Low-Activity Waste (LAW) Facility at the Hanford Waste Treatment and Immobilization Plant. The pretreatment is performed by the Tank Side Cesium Removal (TSCR) system, which prepares LAW for vitrification by removing solids and cesium-137. The solid removal step is conducted by dead-end filtration are operated in parallel: one is actively processing while the other is flushed and put on standby. Thus, the filters periodically restart from a flushed state when the system swaps between them.

Significant transient behavior has not been observed in the deployed TSCR system to date; however, test data collected at multiple scales using the same filter media Time dependence manifests as pressure drop increases required to maintain the target throughput, and in some cases, this occurs shortly after filtration begins. In the future, the TSCR filters may experience feeds with enough solids to experience some transience and anticipating/understanding the behavior would be helpful to guide operations. To that end, test data sets collected with actual and simulated waste are used to investigate fluid and operating parameters that affect transient behavior and examine the impact of filter configuration and scale.

289c Investigation into Na and Cs Activity Coefficients in High Salt Solutions to Support Cs Removal in Hanford Tank Waste

Emily Campbell, Amy Westesen, Reid Peterson

The treatment of Hanford tank waste is one of the most technically challenging environmental cleanup activities for the Department of Energy to date. To expedite the processing of liquid waste stored in the underground tanks in southeastern Washington State and remove the significant dose contributor, Cs-137, ion exchange with crystalline silicotitanate (CST) has been employed as part of the Tank Side Cesium Removal (TSCR) system. The model used to predict Cs exchange onto CST was developed using activity coefficients calculated from the Bromley equation.

A series of batch contact tests from variable Na were conducted to look at the impact of Na concentration on Cs distribution. Experimental distribution ratios were compared to the distribution ratios predicted using three different activity coefficient models 1) commercially available HSC software 2) the Bromley equation and 3) a simplified approach adapted from Marcos-Arroyo et.al. Ultimately the Bromley method underpredicted the effect of ionic strength on the Na activity coefficient, HSC overestimated the impact of ionic strength on the expected performance due to the γ_{Cs+}, but the simplified approach predicted the experimental K_ds quite well in a binary matrix. Expansion of this approach in complex matrices is necessary for application to Hanford tank waste.

289e Density Measurement of Actinide Chloride Molten Salt Systems By Neutron Radiographic Dilatometry

J. Matt Jackson, Scott Parker, Alexander Long, D. Travis Carver, Marisa Monreal

In 2021, Melt point, enthalpy of fusion, and volumetric expansion of single- and multi-component liquid chlorides {NaCl, KCl, LiCl, $MgCl_2$, $CaCl_2$, UCl_3} were measured experimentally. These properties and materials are relevant in applications such as heat transfer, liquid nuclear fuel, and pyrochemical processing. A novel method for density measurement by neutron radiography was shown to produce high-quality data, consistent with reference literature where available, and allowed measurement of some materials for the first time. This method is especially useful in the characterization of sealed sample crucibles, given the ability of neutrons to penetrate the containment and surrounding furnace material. The results of this study are presented within the context of a comprehensive review of the available published data. The purpose of this review is to integrate measurements of the thermophysical properties of liquid chlorides into empirical descriptions of the relationships between composition, temperature, and thermophysical properties. A model for the prediction of the density of mixtures of liquid chlorides is proposed and demonstrated within a case study of the {NaCl + x mol% UCl3} system. These results have been published, and will be presented here.

Recently, $PuCl_3$ bearing samples were prepared at the Plutonium Facility, shipped to the Los Alamos Neutron Science Center (LANSCE), and radiographed with the density via neutron radiography technique described previously. Density data in the form of meniscus heights were recorded as a function of temperature. The results show a linear dependence of density with temperature and a third order polynomial dependence of density with composition as a function of $PuCl_3$ addition in the corresponding halide salt mixture. Additions of $PuCl_3$ result in a significant increase in density of the salt mixture and a slight reduction in the coefficient of volumetric expansion. These results represent a first measurement of the density of plutonium chloride salt mixtures, which once fully analyzed, will lead to a generalized description of the density of such salt systems for arbitrary compositions and temperatures. These plutonium results have yet to be published and will be discussed at the time of the conference.

193a A Constraint on Classical Least Squares for Chemometric Monitoring in Multicomponent Nuclear Waste

Steven Crouse, Rupanjali Gurprasad, Ronald Rousseau, Martha Grover

Advances in spectroscopic equipment and analysis techniques (chemometrics) of vibrational spectroscopy have enabled the use of process analytical technology (PAT) in a variety of applications. A situation often encountered when applying spectroscopy to a continuous monitoring task is that unknown species, or adulterants, can disrupt quantification predicted by PAT. The disruptive behavior of unknown species can lead to poor model robustness and prediction errors, which can lead to misinformed or delayed process decisions. Many methods exist for identifying and removing additional sources of variation that are not included in calibration data. However, existing methods either require estimates of the pure-component spectra of unknown species or rely on multiple time-series measurements to be effective. The method discussed here, a constraint on established Classical Least Squares methods, avoids the requirement of time-series data and the need for pure-component spectra of unknown species, while still proving effective in the case of poorly resolved or overlapping peaks. The applied constraint is that unknown species will have a nonnegative spectral contribution. All species obeying the Beer-Lambert law cannot physically have negative spectral contributions. A constrained Classical Least Squares problem produces a more physically realistic solution than an unconstrained Classical Least Squares solution in cases where spectra closely follow the linearity of the Beer-Lambert Law and unknown species are present. This work focuses on applying this constrained Classical Least Squares approach to multicomponent streams encountered at the Hanford Site in Washington State.

The effective utilization of data has become more important as processes are providing more information with Industry 4.0 initiatives [1]. One application where data utilization has the potential for process improvements is the Hanford nuclear waste vitrification process plant in Washington State. This site removes water and radionuclides from radioactive waste before vitrifying the waste for stable long-term storage. However, the vitrification process is expected to have inconsistent feed stream compositions, leading to the need for real-time process monitoring. A constrained Classical Least Squares approach has the potential to strengthen the robustness of real-time chemometric models monitoring the Hanford site.

Two on-line sensors are investigated in this study: Raman spectroscopy and Attenuated Total Reflectance – Fourier Transform Infrared (ATR-FTIR) spectroscopy. Both are vibrational spectroscopies that can benefit from a constrained Classical Least Squares approach to quantification in complex mixtures. ATR-FTIR derives its linearity from the Beer-Lambert Law, while Raman scattering is analogously linear under certain conditions [2]. Using physical relationships that suggest linearity is the theoretical underpinning of quantification models such as Classical Least Squares. The unknown or adulterating species are also assumed to obey the Beer-Lambert Law and to have nonnegative spectral contributions.

For this study, data were both created computationally and measured experimentally for investigating the efficacy of the proposed constraint on Classical Least Squares. Experimental data have been collected using measurements of anions relevant for Hanford processing: nitrate, nitrite, and sulfate at concentrations representative of the Hanford site. Additional anions unknown to the model calibration are included in the testing experiments: carbonate, oxalate, acetate, and phosphate. The goal of the constrained Classical Least Squares preprocessing is to improve quantification of the calibrated species in the presence of multicomponent mixtures.

Constrained Classical Least Squares is compared to other preprocessing methods in this work: Principal Component Analysis, an artificial neural network autoencoder, Blind Source Separation methods, and Spectra Residual Augmented Classical Least Squares [3]. It is shown that constrained Classical Least Squares outperforms these other methods when there are unknown species present, while also only requiring a single spectrum to operate.

This work has the potential to enable robust real-time monitoring of multicomponent streams like those present at the Hanford site. Nuclear waste processing remains an ongoing environmental, engineering, and political challenge. The results discussed in this work may be a step towards a faster and safer disposal timeline for the Hanford Site.

[1] I. A. Udugama *et al.*, "The Role of Big Data in Industrial (Bio)chemical Process Operations," *Ind. Eng. Chem. Res.*, vol. 59, no. 34, pp. 15283–15297, 2020, doi: 10.1021/acs.iecr.0c01872.

[2] R. L. McCreery, *Raman Spectroscopy for Chemical Analysis*, vol. 157. 2000.

[3] D. M. Haaland and D. K. Melgaard, "New augmented classical least squares methods for improved quantitative spectral analyses," *Vib. Spectrosc.*, vol. 29, no. 1–2, pp. 171–175, 2002, doi: 10.1016/S0924-2031(01)00199-0.

193b Investigating Air Thermal-Hydraulic Behavior Under Natural Convection By Non-Uniform Heating in a Plenum-to-Plenum Facility

Zeyad Zeitoun, Ahmed Jasim, Mahmoud M. Taha, Muthanna Al-Dahhan

Obtaining precise results and new benchmark data for studying the gas thermal behavior flowing within the very high temperature gas cooled reactors (VHTGRs) significantly affects the design and safety operations of these reactors. The two candidates classified as VHTGRs are prismatic modular reactors (PMRs) and pebble bed modular reactors (PBMRs). Therefore, a unique Plenum-to-Plenum facility (P2PF) has been designed and manufactured as a distinguished outstanding representation for the gas flow in the prismatic modular reactors (PMRs) core during loss of flow accidents (LOFA). This facility consists of two plenums connected by vertical dual channels in which one channel is heated and the other channel is cooled to create natural circulation between the two plenums. The objective of the current study is to understand and visualize the air thermal behavior at different axial positions along the hot channel for various heating profiles by integrating advanced measurement techniques. Results show that the air temperature and velocity along the heated channel depends on the hot channel surface temperature distribution. Moreover, flow reversal (back-mixing) is observed at the top of the hot channel leading to the formation of eddies and turbulence which significantly affect the air thermal behavior. The obtained findings will provide valuable data for further investigations by computational fluid dynamics (CFD).

193c Employing Inline Raman Spectroscopy to Determine Solid Properties in Slurries from Solution Phase Attenuation

Rupanjali Gurprasad, Steven Crouse, Ronald Rousseau, Martha Grover

Multiphase and multicomponent radioactive waste at the Hanford site in Washington State poses a severe environmental concern for current and future generations.[1,2] Of the 56 million gallons of legacy nuclear waste at Hanford, approximately 1 million gallons have leaked from underground tanks. To address this environmental issue, the United States Department of Energy is immobilizing the waste into glass for long term storage. The waste still in the underground tanks will be separated into Low Activity Waste (LAW) and High-Level Waste (HLW) before being vitrified into borosilicate glass via the addition of glass-forming chemicals (GFCs).

The addition of solid GFCs into the waste streams result in highly dense slurries. For HLW processing at Hanford, the currently planned process is designed for 20 wt% insoluble solids.[3] For the LAW process, the addition of glass forming chemicals will create a slurry with roughly 22–33 wt% insoluble solids. The physical properties of the added solids, such as concentration and particle size, influence slurry properties (rheology and settling rate), as well as system performance (mixing and pipeline transfer).[4] As a result, slurry density (solid concentration) and particle size are important process parameters for efficient HLW processing. In this study, *in-situ* Raman spectroscopy has been employed as a process analytical tool (PAT) to quantify the concentration of added solids (slurry density) to support the efforts of the real-time inline monitoring (RTIM) program at Hanford. An *in-situ* focused beam reflectance measurement (FBRM) probe was also used to obtain the chord length of the added solids.

The present work focuses on a specific vessel in the waste processing flowsheet at Hanford, namely the melter feed preparation vessel (MFPV). The MFPV represents the point in the process flowsheet where nuclear waste slurries are generated following the addition of solid GFCs to the waste stream for both LAW and HLW processing. The solution phase in the current investigation was a 3 m sodium hydroxide solution (pH>13) containing 1.3 m sodium nitrate (dominant anion in Hanford waste). By adding varied amounts of solid GFC mixtures (developed by Savannah River National Laboratory) to the nitrate solution, waste slurries with varying solid contents were simulated. Solids were added in batches to generate slurries with up to 600 g/L solids (35 wt%). The changes in the Raman intensity of the nitrate anion due to the addition of solids was continuously monitored by tracking the nitrate peak at 1050 cm^{-1}.

The addition of solids was observed to decrease the Raman intensity (attenuate the signal) of the nitrate peak in solution (Fig. 1a). The amount of attenuation was found to be a function of the solid concentration (slurry density, Fig. 1b). A power law function was used to model the attenuation behavior of nitrate anion in the presence of different solid GFC mixtures. The inset in Fig. 1b compares the experimentally measured Raman attenuated intensities with the attenuated intensities predicted by the power law model. The parity plot (inset of Fig. 1b) shows that Raman spectroscopy measurements can be used to extract information about the solid concentrations in slurries by measuring the attenuation of the Raman intensity of a solution phase species (nitrate anion).

The effect of particle size on the attenuation behavior was also investigated by adding silica (main constituent of the glass-forming chemicals) of different particle size distributions to the nitrate solution. The attenuation behavior of nitrate in the presence of two different silica is shown in Fig. 1c. The chord length distribution (CLD) of these solids determined using FBRM (inset in Fig. 1c) verified that the size distribution of these particles was different. Larger silica particles (~30 μm) were found to attenuate the signal the least, whereas smaller silica particles (~15 μm) attenuated the signal the most per gram of silica. This suggests that, in addition to slurry density, the rate of Raman attenuation can be used to extract additional physical properties of the slurries such as particle size.

Traditionally, inline Raman spectroscopy has been employed for quantification of multicomponent systems in food, mining, pharmaceutical and nuclear industries. The results presented in this work highlight the robustness and ability of inline Raman spectroscopy to additionally quantify other properties such as slurry density and particle size of the solids. These physical properties are specifically relevant for efficient HLW processing at Hanford.

Bibliography:

(1) Kelly, S. E. A Joule-Heated Melter Technology for the Treatment and Immobilization of Low-Activity Waste. *Washingt. River Prot. Solut. Rep. RPP-48935* **2011**, *767* (Rev. 0).

(2) Pegg, I. L. Turning Nuclear Waste into Glass. *Phys. Today* **2015**, *68* (2), 33–39.

(3) Goel, A.; McCloy, J. S.; Pokorny, R.; Kruger, A. A. Challenges with Vitrification of Hanford High-Level Waste (HLW) to Borosilicate Glass – An Overview. *J. Non-Crystalline Solids X* **2019**, *4*, 100033.

(4) Wells, B.; Gauglitz, P.; Mahoney, L.; Fountain, M. *Technical Gaps in Hanford High-Level Waste Solids Settling Behavior and Settling Time Evaluation for Direct Feed High-Level Waste (DFHLW) Operations*; 2020.

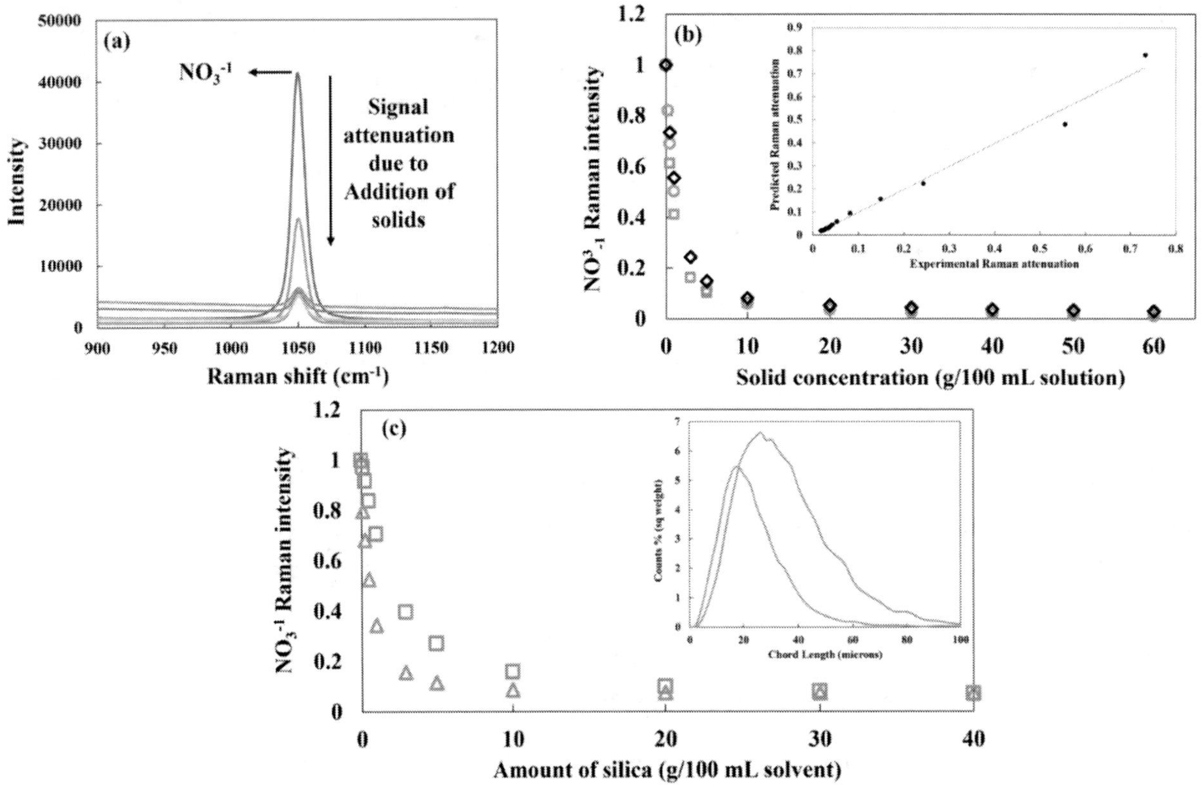

Hanford Sludge Washing for Aluminum and Phosphate Dissolution Duplicate

Amy Westesen, Abigail Robb, Nicholas Cappella, Christian Alvarez, Reid Peterson, Nicholas Machara

The nuclear waste at the Department of Energy's (DOE) Hanford site represents one of the single largest environmental cleanup efforts currently underway. Large volumes of highly radioactive waste, generated during past plutonium production operations, are stored in underground tanks on the Hanford site. Effective disposition of this waste will require processing the sludge to immobilize the waste via vitrification into a stable glass form for long-term disposal. Conceptual flowsheets have proposed settling of the solids in process vessels as a primary step in providing sludge to the vitrification facility. Wastes with high concentrations of aluminum and phosphate have been identified as particularly challenging to process and simultaneously make up the majority of insoluble components in the sludge. This presentation details low temperature dissolution studies of aluminum and phosphate sludge simulants in a variety of wash matrices mimicking potential retrieval operations. Slurry compositions were prepared to match difficult layers in sludge tank AN-106 in order to bound high phosphate/high aluminum operating conditions experienced during processing this tank. Empirical testing data were compared to modeled solubility limits to determine dissolution rates under various wash conditions as well as demonstrate sludge volume reduction capabilities. Removal of these components can significantly reduce (2x) the volume of high-level waste glass produced, decrease operating costs, and improve processing performance.

193d Reduced Temperature Crystalline Silicotitanate Ion Exchange Performance on Cesium Removal in Various Dilutions of Hanford Tank Waste.

Christian Alvarez, Amy Westesen, Emily Campbell, Andrew Carney, Truc Trang-Le, Reid Peterson, Matt Landon

Treatment of AP tank farm supernate to produce immobilized low-activity waste has been initialized at the Hanford Site with the deployment of the Tank-Side Cesium Removal (TSCR) system. Suspended solids and cesium are removed through the utilization of dead-end filtration and crystalline silicotitanate (CST) ion exchange unit operations, respectively. The aqueous waste is fed directly through TSCR from underground storage tanks in preparation for vitrification and long-term storage. Washington River Protection Solutions, which manages the Hanford tank farms and TSCR, requested Pacific Northwest National Laboratory perform bench-scale tests analogous to the full-scale TSCR filtration and ion exchange processes at 16 °C. Current TSCR operations prepare the waste for processing by diluting the highly caustic (>8 M Na) streams with Columbia River water to target a sodium molarity of 5-6 M Na. As part of an effort to reduce river water usage, tests using 5.5 M Na and 7 M Na AP-105 waste were performed using TSCR-prototypic operating conditions to compare CST performance on Cs removal. Each test was performed in a lead-lag column format with a polish column to be added if Cs-137 effluent breakthrough exceeded the waste acceptance criteria (WAC) of 3.18E-5 Ci ^{137}Cs/mol Na. Significant improvement in system performance was observed when processing the diluted (5.5 M Na) AP-105 waste supernate. Testing suggested the throughput capacity of the ion exchange system is increased by 10-20% when the feed is diluted to 5.5 M Na rather than 7 M Na when normalized for Na throughput of the system. The increased performance of the ion exchange system in dilute feed streams can reduce the amount of CST (i.e. number of columns) needed thus reducing the associated operational costs for processing the waste.

193e Bubble Dynamics in Pressurized Water Small Modular Reactor Core Under Loss of Coolant Accident Conditions

Saud Aldawood, Ahmed Jasim, Zeyad Zeitoun, Hayder Al-Naseri, Muhna Alshammari, Al-Dahhan Muthanna

The present investigation aimed to design and construct a downscaled test facility that simulates the bubble dynamics and void fraction in Pressurized Water Small Modular Reactor (PWSMR) to obtain detailed data on the core bubble dynamics and void fractions under loss of coolant scenarios using advanced measurement techniques, to advance the knowledge of fluid dynamics of the pressurized water reactors. The presently available test data lacks the requisite resolution for validating computational fluid dynamics (CFD) and benchmarking tests. To address this research gap, a test section was fabricated with a transparent square wall of 5x5 rods and distinctive grid spacers, utilizing cutting-edge 3D printing techniques. The test facility employed water and air as working fluids, and an innovative 4-point optical fiber probe and data acquisition system were developed to measure bubble characteristics such as local gas holdup, bubble chord length, specific interfacial area, bubble velocity, bubble passage frequency, and their distribution and time series. The experiments were performed at various liquid various gas and liquid flow rates covering bubbly and churn turbulent flow regimes. The acquired data provides valuable information for validating CFD and assessing the distribution of bubbles.

193f Examining the Radiation Stability of Chromatographic Materials

Erin Bertelsen

Radiation stability is a critical factor in the development and application of materials in various fields, such as nuclear energy and medicine. Understanding the effects of ionizing radiation on materials is crucial for designing more radiation-resistant materials, including those used for the separation and purification of radioactive analytes in extraction chromatography.

In this study, the radiation stability of support materials functionalized with bis-(2-ethylhexyl)phosphoric acid (HDEHP) was studied by exposing them to a ~100 kCi ^{60}Co gamma source. The support materials were in contact with an aqueous nitric acid solution of varying concentrations (0.1–1.0 M) and received doses ranging from 10–500 kGy. The extraction chromatographic materials were tested for europium uptake via batch extraction. The extraction performance was compared for three different types of materials: diatomaceous earth, carbon black, and polymeric resin. Additionally, europium uptake was evaluated using nonirradiated and irradiated solutions without a support material (i.e., solvent extraction) to understand the role that support materials play in radiation resistance.

This study aims to elucidate the relationship between support materials and their radiation tolerance, and to offer insight into the design and development of radiation-resistant materials for extraction chromatography. The results of this study will be analyzed to gain better understanding of the radiation stability of support materials and how it impacts their effectiveness in extraction chromatography. These finding will have important practical applications in the development of radiation-resistant materials for the separation and purification of radionuclides.

193g Design and Development of a Radioisotope-Based Technique for Investigating Pebble Flow Characteristics in Cold-Flow Experimental Pebble Bed Reactors

Ahmed Jasim, Jihane Mendil, Saud Aldawood, Haider Al-Rubaye, Zeyad Zeitoun, Al-Dahhan Muthanna

Challenges are posed to investigating the pebble flow dynamics using traditional optical velocimetry methods due to the dense and opaque characteristics of pebble-bed nuclear reactors. As a result, there are only a limited number of experimental studies on pebble flow dynamics in the literature. In this study, a radioisotope-based residence time distribution (RTD) measurement technique was developed to investigate pebble flow in a scaled-down cold-flow experimental pebble bed setup with one pebble at a time discharge rate and continuous pebble recirculation features using 6.0 cm graphite pebbles for the first time. The developed RTD technique uses a set of five collimated Sodium Iodide (NaI) detectors to track the photon counts of a radioactive tracer pebble as it flows down the bed and exits the reactor. The tracer pebble was prepared in-house by tagging one of the pebbles with a Co-60 radioisotope particle of 500 μm in diameter and of an initial activity of 500 μCi. The overall residence time was found to increase as the radius of the initial seeding position increased from the center towards the wall, but with a higher increase rate near the wall compared to the center region. The zonal residence time findings indicated a uniform radial profile of the residence time in the upper section of the bed, except near the wall. However, non-uniformity in the radial profile increased in the lower sections. The cone region had the highest difference in the zonal residence time, with a 200% increase in the tracer's residence time when seeded at the wall ($r/R=0.88$) compared to the center point ($r/R=0.0$). The designs of the experimental setup and the developed technique, including the radioisotope tracer pebble preparation, are described and presented, and the obtained results are analyzed and compared with the literature.

193h Developing Robust and Rapid Methodology of Image Processing System for Individual Ex-Core TRISO-Fueled Pebble Recognition

Jihane Mendil, Roshan Kenia, Ahmed Jasim, Zhaozheng Yin, Al-Dahhan Muthanna, Zeyad Zeitoun

As the world works toward satisfying its electricity demands, nuclear power will play a significant part in the overall mix of energy sources utilized. Future fossil fuel depletion, growing concerns about emissions, the increasing global warming, expanding economies, and the need for sustainable growth and boom ensures that the use of nuclear energy will continue to be important and required in the years to come.

The generation of nuclear power plants changed since its conception to become four distinct generations. Nuclear power plants were subject to continuous development effort that resulted, up to the current day, in four generations. Fourth generation (Gen IV) nuclear reactors are being designed and investigated worldwide. Six innovative designs were proposed as Gen IV candidates for next-generation nuclear plants (NGNPs). The primary advancement of the proposed Gen IV nuclear reactors, compared to older generation nuclear plants, is improved nuclear protection and security, enhanced thermal efficiency, danger-unfastened proliferation (higher resistance to the proliferation of fissile materials), minimal generation of radioactive waste and their safe control, as well as low-cost design. All of these characteristics yield construct cost reduction and operation of NGNPs. Due to the high thermal performance of these reactors, they can be utilized to generate either electricity, in addition to heat generation for usage in industrial settings.

The high temperature gas-cooled reactor, which is one of the six candidates for Gen IV NGNPs, is considered as a promising option for Generation IV advanced reactors. Interest was aroused by the various advantages of this new technology, which include their modularity, enhanced safety, broad applications ranges, and short construction periods. The TRISO-fueled pebbles used in advanced gas-cooled and salt-cooled reactors have a diameter of 6 cm and are made up of an inner fuel zone with a radius of 2.5 cm, which shelters the TRISO nuclear fuel particles and an outside graphite shell with a thickness of 0.5 cm. To measure the burnup, these pebbles are pulled out of the outlet at the base of the nuclear reactor's core one at a time. If their burnup level falls below the threshold, they are sent back to the reactor. It's crucial to prevent excessive burnup accumulation by ensuring that these pebbles are not kept in the core for a prolonged period. To determine whether a pebble has unexpectedly remained inside the reactor core for a long time and to figure out the cause of the excessive burnup accumulation, it is crucial to tag and identify each pebble inside the reactor core. This article provided and verified a novel, robust bullet-time tagging, and tracking system based on computer vision techniques that have been investigated for tagging and tracking each pebble to satisfy the requirement for identifying the residence time of each pebble in the reactor's core. Six spots on the surface of the graphite spheres were imprinted with a three-digit ID number at random using Ultra-High Temperature Ceramic (UHTC) printing. A deep learning-based object detection model bookkept the residence time of each pebble after passing through the bullet-time photography system that can capture the details of each pebble's surface. Once a certain pebble comes out of the pebble bed and passes through the bullet-time photography system again, it will immediately be sought out from the reference library to record the transit or residence time.

193i A Comprehensive Study of Dissolution Methods Utilized to Analyze Insoluble Solids Found in Nuclear Waste: The Cold Chemical Dissolution Method

Dylan Rodene, Madison Hsieh, John M. Pareizs, Stephen Harris, Chuck Coleman

This study focuses on the dissolution of insoluble solids that are a result of spent nuclear fuel and nuclear waste at the Defense Waste Processing Facility (DWPF). The dissolution of insoluble solids is of importance to the analytical qualification of macro batches of nuclear waste, which is necessary to inform and meet processing criteria. For this study, the insoluble solids of interest are that which are found in alkaline adjusted waste tanks and may also be referred to as sludge herein. Elemental analyses of sludge receipt and adjustment tank (SRAT) Receipt and SRAT Product process control samples are studied herein for sludge batch 10. These samples are acquired from a procedure that is performed at DWPF where insoluble solids in waste tanks are blended and processed into vitrified glass (which is identified as a safe long-term waste-form for nuclear waste).

The nuclear waste in DWPF waste tanks is unique in that it contains a wide range of both scarce and abundant elements and chemical species. In order to analyze the wide array of elements and chemical species, three digestion methods were studied. The three methods included a sodium peroxide fusion (PF), a hot aqua regia (AR), and a cold chemical (CC) digestion method. The CC digestion is similar to the DWPF CC method and involved digesting DWPF SRAT samples with high concentrations of HF-HNO_3 acid at room temperature. All samples were studied in quadruplicates within a shielded cell facility at SRNL. Furthermore, this study aids in demonstrating the robustness of various digestion methods by studying insoluble solids that contains a wide gambit of elements, a result of fission products, safety chemicals, corrosion control, cladding materials, etc.

This work provides a statistical analysis and comparison of elemental analysis data of SRAT Receipt and SRAT Product samples generated during an SB10 demonstration. Once digestion was performed, the methods were quantified primarily by inductively coupled plasma – optical emission spectroscopy (ICP-OES). Statistical determinations and equivalence results were based upon Levene, ANOVA, and Welch tests for a total of 19 elements based on elemental analyses. Elemental analyses performed after the CC method did not reveal any major biases for any of the primary elements ($\geq$0.5 wt.% dry solids basis) in the sludge when compared with analyses obtained following dissolution of hot AR or PF methods (even with possible residual solids and fluoride-based precipitates present for the CC method). However, this study did reveal some minor biases and also indicated that no single method is perfect for determining all of the elements, where some elements cannot be considered for specific methods.

The CC elemental analyses agreed with the AR and PF methods well enough that it should be adequate for routine process control analyses in the DWPF. As a result, the CC method can be used to provide DWPF process control. Furthermore, the CC method is implemented as a faster and more convenient sample preparation method than that of the tandem sample preparation methods that employ heating (specifically, hot AR and PF methods performed at 115 and 675 °C, respectively). However, since no heating is utilized for the CC method, it is considered to be a less rigorous method than those based on heating and should be reviewed periodically for its effectiveness in the DWPF process.

AUTHOR INDEX